FLORA OF TROPICAL EAST AFRICA

BUXACEAE

B. Verdcourt

(East African Herbarium)

Trees, shrubs or rarely herbs. Leaves evergreen, simple, alternate or opposite, often coriaceous, exstipulate. Flowers unisexual, monoecious or dioecious (rarely with a few hermaphrodite ones), in spikes, fascicles or cymes. Sepals 0–4, imbricate. Petals absent. Stamens 4–6, more rarely 7–10*, opposite the sepals or with two pairs opposite the inner sepals ; the male flowers often have a rudimentary ovary. Ovary superior, 3-locular ; styles 3, entire or grooved ; ovules 1–2 per loculus, pendulous. Fruit a capsule or a drupe. Seeds black and shining ; endosperm fleshy.

Two genera occur in Africa, *Buxus* L. and *Notobuxus* Oliv. but only the latter occurs in our area.

NOTOBUXUS

Oliv. in Hook., Ic. Pl. 14 : 78, t. 1400 (1882) ; Hutch. in K.B. 1912 : 55 (1912) ; Phillips in Journ. S. Afr. Bot. 9 : 138 (1943)

Macropodandra Gilg in E.J. 28 : 114 (1899)

Small trees or shrubs. Leaves opposite, entire, chartaceous or coriaceous. Flowers subfasciculate or in short cymes, consisting of a solitary sessile or subsessile ♀ flower subtended by two bracts, surrounded by a few pedicellate, bracteate ♂ flowers ; or rarely the entire inflorescence composed of ♂ flowers**. Male flowers : sepals 4, obovate or orbicular, the outer narrower and subcucculate ; stamens usually 6, in two series, the outer series of 2, each opposite an outer sepal, the inner series of 4, each pair opposite an inner sepal ; anthers sessile ; rudimentary ovary absent. Female flowers : sepals 4 ; styles 3, thick, grooved and recurved ; ovules 2 per loculus. Fruit capsular, dehiscing loculicidally ; valves each with 2 apical horn-like projections. Seeds keeled.

A small genus of 7 species occurring in tropical and South Africa and also in Madagascar.

N. obtusifolius *Mildbr.* in N.B.G.B. 12 : 710 (1935) ; T.T.C.L. : 87 (1949) ; Verdc. in K.B. 10 : 598 (1956). Type : Tanganyika, Lindi District, Mlinguru, *Schlieben* 5818 (B, holo., BR, iso. !, EA, K, photo-iso. !)

A small shrub, 1–1·2 m. tall or small tree, 6 m. tall ; branches grooved, pubescent or glabrous ; bark corky. Leaves oblong-elliptic, obovate-oblong or ovate, cuneate at the base, narrowed to an obtuse or emarginate apex, coriaceous, deep green above, paler beneath, glabrous or pubescent on the midrib beneath, 1·6–7·6 cm. long and 0·75–3·4 cm. wide, with the venation raised on both surfaces or impressed above ; petiole thickened,

---

* I have dissected flowers of *N. natalensis* Oliv. having 7, 8, 9 or 10 anthers.
** I have found this condition in *N. natalensis* Oliv.

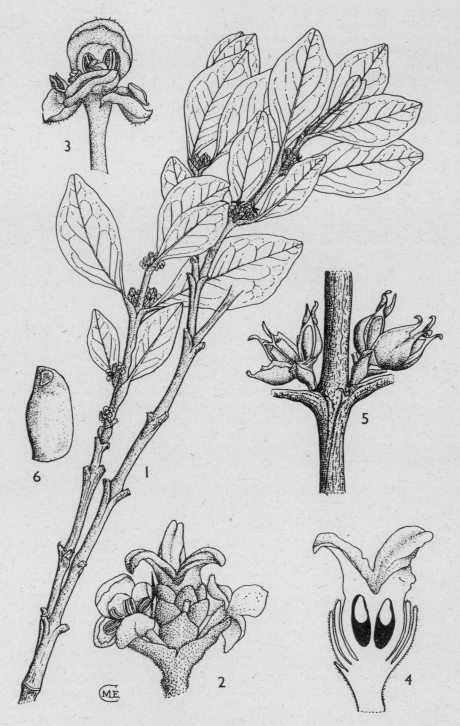

FIG. 1. *NOTOBUXUS OBTUSIFOLIUS*— **1**, flowering branch, × 1 ; **2**, inflorescence, × 6 ; **3**, ♂ flower, × 8 ; **4**, ♀ flower in L.S., × 8 ; **5**, fruits, after dehiscence, × 2 ; **6**, seed, × 6. 1–4, from *Greenway* 4138 ; 5, 6, from *Rawlins* 765.

2–5 mm. long. Flowers green, in reduced axillary inflorescences consisting of one terminal ♀ flower with four ♂ flowers beneath it, one of which may be somewhat undeveloped ; peduncles short, flattened.   Anthers ovoid-oblong, about 1 mm. long. Capsule ovoid, black, 7 mm. long, dividing into three valves. Seeds oblong-ovoid or trigonous, about 4 mm. long and 2 mm. wide. Fig. 1.

KENYA. Kwale District : Gazi, 23 Febr. 1959, *Moomaw* 1432 ; Kilifi District : North Giriama Reserve, N. of Sabaki River near Adu, Jan. 1937, *Dale* 3631 ! ; & Arabuko-Sokoke forest, 1·6 km. from Arabuko Sawmills, June 1959, *Rawlins* 765 !

TANGANYIKA. Lushoto District : East Usambara Mts., Kijango, 27 Oct. 1935, *Greenway* 4138 ! ; Lindi District : Mlinguru, 20 km. south of Lindi, 2 Feb. 1935, *Schlieben* 5818 !

DISTR. **K7** ; **T3**, 8 ; not known elsewhere

HAB. Evergreen *Euphorbia-Brachylaena* thicket, *Cynometra-Brachylaena* forest or dry scrubby forest on red soils ; 280–450 m.

NOTE. At first sight, from herbarium specimens, *N. obtusifolius* appears almost identical in facies with *N. natalensis* Oliv. which occurs in Natal and such a distribution would not be at all unexpected. *N. natalensis* has, however, two, rarely three, larger male flowers per inflorescence whereas *N. obtusifolius* has four smaller ones ; the horn-like projections on the capsule are also longer in *N. natalensis*, and the anthers are 2·5 mm. long.

INDEX TO BUXACEAE